Biology Science Fair Projects
FOR
DUMMIES
MINI EDITION

by Maxine Levaren

John Wiley & Sons, Inc.

Biology Science Fair Projects For Dummies®, Mini Edition

Published by
John Wiley & Sons, Inc.
111 River St.
Hoboken, NJ 07030-5774
www.wiley.com

Copyright © 2013 by John Wiley & Sons, Inc., Hoboken, New Jersey

Published by John Wiley & Sons, Inc., Hoboken, New Jersey

Published simultaneously in Canada

No part of this publication may be reproduced, stored in a retrieval system or transmitted in any form or by any means, electronic, mechanical, photocopying, recording, scanning or otherwise, except as permitted under Sections 107 or 108 of the 1976 United States Copyright Act, without the prior written permission of the Publisher. Requests to the Publisher for permission should be addressed to the Permissions Department, John Wiley & Sons, Inc., 111 River Street, Hoboken, NJ 07030, (201) 748-6011, fax (201) 748-6008, or online at http://www.wiley.com/go/permissions.

Trademarks: Wiley, the Wiley logo, For Dummies, the Dummies Man logo, A Reference for the Rest of Us!, The Dummies Way, Dummies Daily, The Fun and Easy Way, Dummies.com, Making Everything Easier, and related trade dress are trademarks or registered trademarks of John Wiley & Sons, Inc. and/or its affiliates in the United States and other countries, and may not be used without written permission. All other trademarks are the property of their respective owners. John Wiley & Sons, Inc., is not associated with any product or vendor mentioned in this book.

LIMIT OF LIABILITY/DISCLAIMER OF WARRANTY: THE PUBLISHER AND THE AUTHOR MAKE NO REPRESENTATIONS OR WARRANTIES WITH RESPECT TO THE ACCURACY OR COMPLETENESS OF THE CONTENTS OF THIS WORK AND SPECIFICALLY DISCLAIM ALL WARRANTIES, INCLUDING WITHOUT LIMITATION WARRANTIES OF FITNESS FOR A PARTICULAR PURPOSE. NO WARRANTY MAY BE CREATED OR EXTENDED BY SALES OR PROMOTIONAL MATERIALS. THE ADVICE AND STRATEGIES CONTAINED HEREIN MAY NOT BE SUITABLE FOR EVERY SITUATION. THIS WORK IS SOLD WITH THE UNDERSTANDING THAT THE PUBLISHER IS NOT ENGAGED IN RENDERING LEGAL, ACCOUNTING, OR OTHER PROFESSIONAL SERVICES. IF PROFESSIONAL ASSISTANCE IS REQUIRED, THE SERVICES OF A COMPETENT PROFESSIONAL PERSON SHOULD BE SOUGHT. NEITHER THE PUBLISHER NOR THE AUTHOR SHALL BE LIABLE FOR DAMAGES ARISING HEREFROM. THE FACT THAT AN ORGANIZATION OR WEBSITE IS REFERRED TO IN THIS WORK AS A CITATION AND/OR A POTENTIAL SOURCE OF FURTHER INFORMATION DOES NOT MEAN THAT THE AUTHOR OR THE PUBLISHER ENDORSES THE INFORMATION THE ORGANIZATION OR WEBSITE MAY PROVIDE OR RECOMMENDATIONS IT MAY MAKE. FURTHER, READERS SHOULD BE AWARE THAT INTERNET WEBSITES LISTED IN THIS WORK MAY HAVE CHANGED OR DISAPPEARED BETWEEN WHEN THIS WORK WAS WRITTEN AND WHEN IT IS READ.

For general information on our other products and services, please contact our Customer Care Department within the U.S. at 877-762-2974, outside the U.S. at 317-572-3993, or fax 317-572-4002.

For technical support, please visit www.wiley.com/techsupport.

Wiley publishes in a variety of print and electronic formats and by print-on-demand. Some material included with standard print versions of this book may not be included in e-books or in print-on-demand. If this book refers to media such as a CD or DVD that is not included in the version you purchased, you may download this material at http://booksupport.wiley.com. For more information about Wiley products, visit www.wiley.com.

ISBN 978-1-118-50060-6 (pbk); ISBN 978-1-118-54482-2 (ebk);
ISBN 978-1-118-54488-4 (ebk); ISBN 978-1-118-54489-1 (ebk)

Manufactured in the United States of America

10 9 8 7 6 5 4 3 2 1

Table of Contents

Introduction **1**
Icons Used in This Book 2
Where to Go from Here 2

Chapter 1: Science Fair Projects 101 .... **3**
Explaining What's Involved 3
How Science Fair Projects Can Help You 4
Finding the Best Project for You 5
Investigating and Reporting Background
   Information 7
Doing Your Project 7
Presenting Your Work 9
Exhibiting at the Science Fair 9

**Chapter 2: Tackling and Categorizing Your
Projects** **11**
Considering Four Ways to Do a Science Project .. 12
Running Down the Official ISEF Project Categories ... 13

**Chapter 3: What's the Big Idea? Finding a
Project Topic** **23**
Using What You're Studying in Class 24
Getting Inspiration from Media Sources 24
Cultivating Ideas from Current Events and Issues ... 25
Parlaying Personal Interests 26
Building on a Previous Project 27
Applying the Acid Test: Can I Really Do
   This Project? 28

Chapter 4: Starting Out with an Easy Project 33
Behavioral and Social Sciences 33
Botany 35
Microbiology 38

Chapter 5: Kicking It Up a Notch: Medium Difficulty Projects 41
The Even-Handed Teacher 41
Hand Sanitizers: Hype or Help? 43

Chapter 6: Taking the Challenge: Harder Projects 47
To Clean or Not to Clean 48
More Biology Challenges 52

Chapter 7: Ten Survival Tips for Parents 53
Supporting Versus Nagging 53
Discovering New Things 54
Making Friends with Your Child 54
Living Your Second Childhood 54
Knowing When to Say No 54
Taking Time for Yourself 55
Staying Centered 55
Getting a Self-Checkup 55
Letting It Go 56
Asking for Help 56

Introduction

Yikes! I have to do a science project! If you've never done one before but have heard horror stories about them, you're probably dreading the whole thing. Contrary to popular belief, doing a science project isn't the 21st century equivalent of medieval torture. True, you may be required to do some hard work, and you may feel at times like you can never get through it, but you'll survive, without any visible scars.

The good news is that you already have a lot of the abilities that you need. Do you like to write? The lab report gives you a place to show off your way with words. If you're an artist in the making, you can have fun creating an attractive project display. If you're persistent, never giving up until you find the answer, research is made for you. And if you're crazy about details, lists, and statistics, I know you can keep excellent, precise records.

This book is a quick and easy introduction to doing a successful science fair project and having some fun in the process. Rather than discussing only projects that involve experiments, I include other types, such as:

- Computer projects that either develop programs to solve a particular problem or analyze system performance.
- Engineering projects that design and build new devices or test existing devices to compare and analyze performance.

✔ Research projects that collect data, either by surveying a specific population or researching current sources, and that mathematically analyze and compare the information.

Icons Used in This Book

Here are the two icons I use and what they signify.

This icon indicates something you need to keep in mind in order to have a successful project.

The bull's-eye marks information that can make your life easier while you're working on your project.

Where to Go from Here

Glance through the Table of Contents and find the chapter or section that flips your switch. If you're trying to figure out what kinds of projects are acceptable, flip to Chapter 2. If you're struggling to come up with a specific idea for your own project, Chapter 3 can help. And if you want to see some examples of easy, moderate, and challenging projects that other students have done, Chapters 4, 5, and 6 are your destination. If you want even more advice on biology science fair projects, check out the full-size version of *Science Fair Projects For Dummies* — simply head to your local bookseller or go to www.dummies.com!

Chapter 1
Science Fair Projects 101

In This Chapter
▶ Getting started without getting stymied
▶ Finding the facts you need
▶ Promoting your work and reaping the rewards

*Y*ou don't have to be a rocket scientist (or even a brain surgeon) to do a successful and fun science fair project, especially if you read this book. In this chapter, I give you a preview of coming attractions, so you can see what a science fair project is all about.

Explaining What's Involved

A *science fair* is an exhibition where students display scientific experiments, computer programs or systems, or engineering designs and tests that they've created. Each project in the fair demonstrates that the students understand the scientific method and can apply it, while having an independent, hands-on learning experience.

Science fairs grew out of teachers' desire to advance interest in scientific knowledge. Now science fairs range from small school exhibitions to one huge international science and engineering fair (the Intel International

Science and Engineering Fair [ISEF]), where students from all over the world show off their work and compete for recognition and valuable prizes.

 The purpose of a science fair is to help you (the student) find out more about a specific topic. Don't think that you need to choose a difficult or complex project to impress the judges. Selecting a project that you can understand, doing it by yourself, and explaining it to the judges at the fair is better.

How Science Fair Projects Can Help You

In this section, I offer a short and relatively painless lecture about why science projects are good for you.

Identifying the practical benefits

Do you like science? Are you good at it? You won't know unless you try, right?

When you do a science experiment in class, the materials come from the supply closet and your teacher or a textbook spells out the procedures. When you work on a science fair project, you call the shots. You decide what your project is about, research your topic, write a paper, and design and execute a project plan.

Besides allowing you to assert your independence, doing a science project also has some more practical benefits. For example, you can

- Collaborate with scientists and engineers
- Hone your mathematical skills
- Try out new computer hardware and software

- ✓ Use power tools and build your own equipment
- ✓ Create artistic displays
- ✓ Give oral presentations
- ✓ Put something cool on your college application

Getting even more from your project

Here are some other takeaways from doing a successful project:

- ✓ **Broadening your knowledge:** When you do a science project, you're not just working on science. You're actually doing math, art, writing, and maybe even computer science. (You can also find out about psychology, history, or anthropology, depending on your project.)

- ✓ **Finding your hidden talents:** Even if you aren't a budding scientist, you may find that you have other talents, such as writing or research, that can help you during the project and beyond.

- ✓ **Discovering new things about yourself:** Besides discovering what things you like (and dislike), you also can find out how you like to work, be it independently or with a support team.

Finding the Best Project for You

Science fair projects can range from the simplest experiment to a complex project that involves experimentation, observation, and advanced mathematical calculations. Where you fall on that scale naturally depends on your current knowledge of science, particularly in the subject matter you're dealing with.

Identifying the different types of projects

You can choose from several different types of projects. Chapter 2 has a full explanation, but here's a quick look:

- In an *experiment,* you conduct a test to prove that a certain assumption is true.

- With a *research* project, you gather data, either by doing research or conducting surveys, and then mathematically analyze that data.

- In an *engineering* project, you either design and build a new device that performs a specific function or you test and analyze the performance and capabilities of an existing device.

- With a *computer science* project, you either develop one or more computer programs to perform a specific task or you test and analyze computer performance.

Choosing among the categories

Science fair projects are also broken down into a number of categories in the biological and physical sciences (see Chapter 2). Check out the official ISEF website for examples of the types of projects in each category: www.societyforscience.org/page.aspx?pid=470.

Making the project suit you

The winning formula for a project is selecting something that you like, that you have the skills to do, and that stretches your abilities. Check out Chapter 3 for some hints on how to find a project that suits you to a "T." Also check out www.sciencebuddies.org and use the survey wizard to find a subject suited to you.

Investigating and Reporting Background Information

The first steps of a project are research and writing:

- **Research, research, research:** First you have to find all the information you need about your topic. From there, you can pick out the facts that directly apply to your topic and then organize them in preparation to write your research paper.

- **Tackling your paper:** You then write your research paper to summarize what you've found out about your topic. The information in your paper serves as a springboard for the project that you'll be doing.

Doing Your Project

Actually doing your project is the heart of the matter:

- **Starting with the scientific method:** Applying the scientific method is crucial. How you use the scientific method varies depending on what type of project you're doing.

  Briefly, the *scientific method* consists of stating a problem or asking a question, and then making an educated guess about the answer or solution, which is called the *hypothesis*. Most experiments manage *subjects* (things to be tested), arranged in *experimental groups* (subjects with changes applied) and *control groups* (subjects with no changes applied). *Variables* are the conditions that are changed and measured; *controls* are the factors that remain constant.

- ✔ **Organizing your materials:** An integral part of your project is a list of materials that you use. You may buy, borrow, or build the things you need.

- ✔ **Explaining your procedures:** Next, you need to explain exactly how you plan to do the project. You may use a narrative in non-regional competitions, but procedures are typically compiled in a sequential, numbered list. That method makes it easier for a teacher or a judge to understand what you're doing.

- ✔ **Keeping good records and noting irregularities:** Whenever you work on your project, you need to keep timely, accurate, and complete records of everything that happens, including unusual things such as a gust of wind that knocks over the rack of samples in the backyard, because these events may affect the results and influence your conclusions. Don't worry if your project didn't work as you expected. More importantly, you need to have reliable proof of what occurred.

- ✔ **Compiling results:** In order to make the data meaningful, you have to arrange it in a clear and intelligible way. This may involve doing some calculations, such as taking an average, arranging the data into tables, or graphing the measurements that you've collected.

- ✔ **Making conclusions:** In your conclusions, you analyze and interpret your project's results, which means you compare the results to the hypothesis, your prediction of what would happen. Either your results proved the hypothesis or they didn't support your theory. You also want to consider whether you'd do things differently if you were repeating the project; note any potential improvements in your conclusions as well.

Presenting Your Work

Two key pieces remain — the display and lab report:

- You use your display to get attention. Think of it as a billboard that advertises your project. You want to entice the science fair judges to take a closer look at your work. To do that, the display has to be attractive and easy to understand. Use color, texture, and design to the best advantage.

- After you've piqued their interest, you need to have something of substance to show the judges — your lab report. It contains your research, materials and procedure lists, logs, results, and conclusions.

For more info about presenting your work, check out *Displaying & Presenting Science Fair Projects For Dummies,* Mini Edition (John Wiley & Sons, Inc.).

Exhibiting at the Science Fair

After you transport your project to the science fair, an official directs you to a spot where you can exhibit it. You set up your display board and lab report, arrange any project material that you want to display, and plug in anything electrical.

You can stand there while the judges walk past your display, look through your lab report, and (gasp!) ask you questions about your project. The best thing you can do at this point is to know your subject, relax, smile, and answer the questions.

After the judging is done, meet and talk to other students who are exhibiting at the fair. If special activities have been arranged for exhibitors, take advantage of them.

10

Nearly every science fair I've attended has had an awards ceremony, where awards, ribbons, and prizes are given. Go, root for your classmates, and celebrate a job well done!

Chapter 2

Tackling and Categorizing Your Projects

In This Chapter
- ▶ Identifying the four different types of projects
- ▶ Examining the official ISEF project categories

*A*ll branches of science aim to discover new information, but not all discoveries happen in a lab. Although doing an experiment is likely the first idea that jumps to your mind when thinking about a science project, other ways to do an effective project do exist.

For example, you can do a research project where you collect and analyze data, or if you love computers, you can write a computer program or analyze computer performance. Are you an inventor in the making? Check out the possibilities for doing an engineering project.

In this chapter, I show you how each type of project is constructed. I also list the official Intel International Science and Engineering Fair (ISEF) categories and describe some project titles in each one. If you want to know more about ISEF, check out Chapter 3.

Considering Four Ways to Do a Science Project

The object of all science projects is to answer a question or solve a problem. Another thing all projects have in common is a *hypothesis,* which is an educated guess about the results and conclusions of a project. However, the type of procedure and the kind of data collected differ depending on the type of project:

- **Performing an experiment:** For this type of project, you perform a procedure according to the scientific method in order to see whether a stated hypothesis is supported.

- **Conducting research:** In a research project, you gather information, analyze it, graph it, apply mathematical calculations, and describe what the data shows.

    Doing a research project isn't the same as writing a research paper. To have an effective research project, you must collect all the data and do the analysis yourself.

- **Building or testing a device:** You can do an engineering project in one of two ways. You can

    - Devise something new that solves a specific problem, build it, and test it. An important component of this type of project is applying variations and seeing how they affect the working of the device.

    - Test and compare existing devices or substances.

✔ **Developing a computer program:** Your project can design, code, and test one or more computer programs that solve a specified problem. It helps to vary the computer project design and test its effects.

Running Down the Official ISEF Project Categories

Many categories of projects compete in science fairs. They cover all scientific and engineering disciplines, ranging from botany to zoology and everything in between. If you already have an idea for your project, check out the list of categories in this section to see where yours belongs. If you're clueless, investigating the categories may give you a few good ideas.

In this section, I describe each category and also give examples of projects that have been exhibited in the category.

Keep in mind that a project can sometimes go in one or more categories. For example, if you're testing water purity, does the project belong in chemistry or environmental science? Does a project about eating disorders fit in behavioral and social sciences or medicine and health? When in doubt, your science teacher is your best guide.

Note: When you see the words Phase II, Phase III, or something similar in a project title, it means that the project is a continuation of a prior year's work.

Biological sciences

All the *biological sciences* look at the world of living things. Just as everything on earth is made up of atoms, every living thing is made up of cells. A living thing can be a single cell organism, such as an amoeba, or it can have billions of cells, such as a tree or a human being.

Behavioral and social sciences

Why do people (or animals) do what they do? Projects in the *behavioral and social sciences* category examine psychology, sociology, anthropology, archaeology, *ethnology* (the science of human cultures), linguistics, learned or instinctive animal behavior, learning, perception, urban problems, reading problems, public opinion surveys, and educational testing.

Examples of behavioral and social science projects are:

- Leading polls: Can rewording sentences influence survey results?
- Multitasking: Productive or disruptive?
- How confidence or testing environment affects math scores
- Trying to teach an old dog (or fish) new tricks

Biochemistry

Okay, if all living things are composed of cells, what are cells made of? Cells are made of molecules, which consist of chemical compounds. And that's *biochemistry* — the chemistry of biological organisms.

Biochemistry projects deal with the chemistry of life processes, including molecular biology and genetics, enzymes, photosynthesis, blood chemistry, protein chemistry, food chemistry, and hormones.

Some examples of biochemistry projects are:

- ✔ Cholesterol analysis of various poultry species
- ✔ Bacterial comparison of organic and pasteurized milk
- ✔ Stress tests
- ✔ Effect of soil types on the decomposition rate of buried apples (or other materials)

Botany

Botany is the study of all kinds of plant life. Projects that deal with agriculture, *agronomy* (the study of soil relating to soil management and crop production), horticulture, forestry, plant biorhythms, anatomy, *taxonomy* (the classification of organisms), physiology, pathology, *hydroponics* (the study of plants grown in nutrient solution), *algology* (the study of algae), or *mycology* (the study of fungus) belong here.

Here's a list of some botany project titles:

- ✔ Can calcium levels in plants be raised? Phase III
- ✔ Effects of microwave heating on seed germination
- ✔ Relationships between proteins and food allergens
- ✔ Tissue culture on a genetically modified plant
- ✔ Geotropism: The response of root growth to gravity

Gerontology

Gerontology is the study of the aging process in living organisms. Here are examples of gerontology projects:

- ✔ Effects of age on sensory memory
- ✔ Relationship between metabolic rate and life span

- ✔ Remembered and forgotten: Autobiographical memory
- ✔ Reactions to distractions

Medicine and health

The *medicine and health* field is concerned with diseases, health, and the healing of animals and humans. Medicine, dentistry, pharmacology, veterinary medicine, pathology, ophthalmology, nutrition, sanitation, pediatrics, dermatology, allergies, speech and hearing, and optometry are good candidates for this category.

Here are some examples:

- ✔ Why are sharks resistant to cancer? The unique core structure of the shark P53 gene
- ✔ Comparison of citronella oil extracts as mosquito repellents
- ✔ Diabetes: How different exercises affect blood sugar
- ✔ Should you wear sunglasses while checking your e-mail?

Microbiology

Microbiology is the branch of biology that deals with microorganisms and their effects on other living organisms. This category includes bacteriology, *virology* (study of viruses and viral diseases), *protozoology* (study of *protozoan*, microscopic single-cell animals), fungal and bacterial genetics, and so on.

To put it in simple terms, microbiology projects deal with germs, as shown in the following list of possible project titles:

- Are you in danger of "Superbugs"?
- Antibiotic-resistant bacteria: How common are they?
- Quality tests of pasteurized raw milk
- Pathogenic bacteria on soft-drink cans: Are they harmful?

Zoology

The science of *zoology* includes all studies of animals, including genetics, *ornithology* (the study of birds), *ichthyology* (the study of fish), *herpetology* (the study of reptiles and amphibians), *entomology* (the study of insects), animal ecology, anatomy, paleontology, cellular physiology, animal biorhythms, animal husbandry, *cytology* (the study of the formation, structure, and function of cells), *histology* (the study of the microscopic structure of animal and plant tissues), animal physiology, neurophysiology, and invertebrate biology.

Following are some sample zoology projects:

- Gerbil coat color genetics
- Microstructure of the shark's jaw
- Effect of colony size on resilience in ants
- Effects of melatonin on the cellular processes in selected organisms

Physical sciences

Physical sciences, such as chemistry and physics, analyze and compare energy and nonliving matter. In terms of science projects, computer science, engineering, and mathematics are also physical sciences.

Chemistry

Chemistry deals with the composition, structure, properties, and reactions of different substances. Projects in this category include physical, inorganic, and organic chemistry (other than biochemistry), materials, plastics, fuels, pesticides, metallurgy, and soil chemistry. Projects that analyze fossils, fuels, vitamins, or crystals also belong in this category.

Here are some examples:

- Extract from catnip as mosquito repellent
- Sunscreens: Analyzing active ingredients for UV absorption
- Converting carbon directly into electricity
- Using water as fuel
- Application of eco-friendly dyes on natural fibers

Computer science

Computer science projects include the study and development of computer hardware, software engineering, Internet networking, communications, and graphics, including human interface. Simulations, virtual reality, or computations science (including data structures, encryption, coding, and information theory) are other topics in this category.

Check out some ideas:

- Advanced video animation compression
- Virtual flashcards

- ✔ An improved algorithm for meshing large and complex terrains
- ✔ Modeling global warming

Earth and space sciences

Earth and space sciences projects deal with soil, minerals, petroleum, weather, and the atmosphere. They cover topics in geology, geophysics, physical oceanography, meteorology, atmospheric physics, seismology, geography, *speleology* (the scientific study of caves), mineralogy, topography, optical astronomy, radio astronomy, and astrophysics. If you're analyzing the geological age of fossils, your project also belongs in this category.

Here are some other sample projects:

- ✔ Developing electron spin resonance (ESR) dating for sharks' teeth
- ✔ How solar flares affect Earth's magnetic field
- ✔ Coastal community risk from tsunami waves
- ✔ Do certain jet stream patterns affect tornado formation and location?

Engineering

If you're doing a project that applies scientific methods to manufacturing and testing, then your project belongs in the *engineering* category. This category includes civil, mechanical, aeronautical, chemical, electrical, automotive, marine, and environmental engineering. Photography, sound, heating and refrigeration, transportation, power transmission and generation, electronics, communications, architecture, bioengineering, and lasers can also fit into this category.

Check out some sample project names:

- ✔ The design of a localized positioning system
- ✔ The most effective material for sleeping bag insulation under wet and dry conditions
- ✔ Recycling waste products: kiln dust, slag, and strength in concrete
- ✔ Design of ergonomic desk for students with scoliosis and other conditions of the spinal cord

 Just because you design and build a device doesn't necessarily make it an engineering project. For example, if you had to design and construct a *still* (an apparatus to distill liquids) to do a project that tests methods of *desalinating* (removing salts or minerals) water, the project belongs in chemistry and not in engineering.

Environmental sciences

Any project that deals with the environment on land, in the air, or in the water belongs in the *environmental sciences* category. Environmental science projects deal with the sources and control of pollution, waste disposal, impact studies, environmental alteration (heat, light, irrigation, erosion, and so on), and ecology. This category also includes the effects of pollution on different organisms.

Take a look at some sample projects:

- ✔ Effects of runoff on water quality
- ✔ Cost-effective native plant roadside restoration
- ✔ Bird populations as indicators of gulf coast environmental quality
- ✔ Fibre bag: An alternate for polybags

Mathematics

The science of mathematics deals with logical, numerical, and algebraic principles and systems, including calculus, geometry, abstract algebra, number theory, statistics, complex analysis, probability, *topology* (history of a region influenced by its physical surface), logic, operations research, and other topics in pure and applied mathematics.

Here are a few titles of mathematics projects:

- ✔ Game theory models for Middle Eastern policy
- ✔ Pitch frequencies of Bach's No. 13 invention
- ✔ Winter wonderland: A mathematical analysis of snowflakes
- ✔ Pattern counting on chessboards

Physics

Physics is the science encompassing the theories, principles, and laws governing energy and the effect of energy on matter. Projects that deal with solid state, optics, acoustics, particle, nuclear, atomic, or plasma physics, superconductivity, fluid and gas dynamics, thermodynamics, semiconductors, magnetism, quantum mechanics, and biophysics belong in this category.

Some sample titles of physics projects are:

- Designing for heat resistance
- Microwave energy as an auxiliary source of heat
- Heat transfer and mass change
- Low-cost plasma reactor
- The behavior of light through various substances

Chapter 3

What's the Big Idea? Finding a Project Topic

In This Chapter
▶ Getting ideas from school or the media
▶ Capitalizing on current events and your interests
▶ Consulting the people you know
▶ Improving an old project
▶ Checking it twice: Is my idea doable?

The most important ingredient in a successful science fair project is the right project idea: something that captures your imagination and keeps you interested and motivated. Ideally, the right idea isn't so difficult that it frustrates and discourages you, but it's challenging enough to stretch your abilities and creativity.

Finding a good topic can be the hardest part of the entire science fair experience, but a good idea really isn't that hard to find if you know where to look. In this chapter, I explain ways you can find a great idea for your project and then help you figure out if your idea is possible.

Using What You're Studying in Class

Your first thought may be to get a project idea from your science class. What you're currently studying can be the springboard to a great project topic. Besides, brainstorming for ideas in science class gives you a built-in source of help and information, because your teacher is right there with you.

 But, be sure to do your project on a topic that you already know a little something about. Although learning is one of the main objectives of your science project, now is a really bad time to take a crash course in calculus if elementary algebra is giving you grief.

Your other classes can also inspire you with a science project idea. Even your social studies classes may get your gray cells going. For example, a class discussion on subliminal advertising may lead to an animal behavior experiment that measures the effect of sound on a mouse's appetite. History courses that deal with prehistory may lead to a project dealing with fossils or dating ancient artifacts.

Getting Inspiration from Media Sources

If you look and listen carefully, you can find loads of project ideas in the media. Here are some thoughts:

- **The written word:** You may find something that sparks your curiosity in a newspaper or magazine. Feature articles often discuss areas of science, especially those that are important in your city or state.
- **The virtual word:** You may wonder whether certain things you read online are true, or you may find that a blog or website motivates you to find out more about a certain topic.
- **The visual world:** Movies or television shows, even if they're fiction, can give you ideas for a science project, as can commercials. (If three battery brands all claim to last the longest, which one is telling the truth?)

Cultivating Ideas from Current Events and Issues

Current events can generate great science project ideas. For example, environmental issues (such as climate change) and medical research (such as cancer or other illness studies) are popular sources of project ideas, as are people's reactions to world events or the psychological impact of such events.

Local issues are often springboards to project ideas as well. Look at a West Coast science fair, and you may see projects about ocean water and sea life. A science fair in Louisiana or East Texas may feature projects that are concerned with plant and animal life in the bayous, and students in the Appalachian region may have projects that deal with the effects of coal mining on the environment.

Parlaying Personal Interests

To have a successful project, choose a project that can be fun. Look at your surroundings, your interests, and your concerns, and you may find a project idea staring you in the face. Here are some ideas:

- **Sports, games, and hobbies:** A project may deal with the statistical side of sports, such as whether the home-field advantage makes a difference in team sports. Or it may deal with how to play the game — for example, analyzing the best angle for taking a shot on a goal. However, the majority of these projects, which you may find mainly in the engineering and physics categories, test different types of sporting equipment. For example:

    - Curvature of a baseball
    - Tennis ball bounce
    - Swimsuit drag

    If your interests run more toward games than sports, you may want to focus your project on a mathematical project such as the odds of winning games of chance. If you enjoy paper airplanes, loads of engineering and physics projects analyze speed, lift, and distance.

- **Everyday life:** I bet you didn't know how helpful your house or apartment can be when trying to find a project idea. Start in the kitchen, where personal interests and science come together. For example, you can find lots of project ideas in food, such as the amount of sugar, fats, or acid. Going a step further, you can look at how to ripen fruit faster or keep food fresher longer.

Moving into the workshop, you can find project ideas looking at rust resistance, paint durability, glue strength, and insulation effectiveness. In the bathroom, you find even more ideas, such as how long soap bubbles last, which whitener makes your teeth the whitest, and which deodorant works best. Continue your home-and-garden tour to find out about flame-retardant fabrics, paper towel absorbency, and whether washable markers are really washable.

- **School and social life:** Problems affecting teenagers can inspire many projects. If you want to gather people's opinions via a survey, you may opt for a project that deals with relationships. Or you may structure a project focused on gender issues or the desires to look and feel good.

- **Your future career goals:** Science fair judges love it when you can relate your project to a topic that you plan to pursue in college or later in life.

Building on a Previous Project

Exhibiting at the Intel International Science and Engineering Fair (ISEF) is, for many students, the culmination of a lot of work. (See the sidebar, "The granddaddy of science fairs," in this chapter for more about ISEF.) A required first project turns into a consuming interest, which becomes a project that gets more polished and professional each year.

If you look at a catalog from a previous science fair, you can get lots of ideas. These catalogs list all the projects, arranged by category, along with the name and school of each exhibitor.

> ### The granddaddy of science fairs
>
> The Intel International Science and Engineering Fair (ISEF) is the world's largest science fair. Each year, high school students, teachers, corporate executives, and government officials from all over the world gather to celebrate and reward scientific inquiry. Although Intel is ISEF's main sponsor, many other corporations and institutions fund the fair and donate student awards and prizes.
>
> The ISEF, which is held in a different city each year, also sponsors local and regional fairs throughout the world. The local fairs, which comply with ISEF rules and guidelines, send their first-place award winners to exhibit their work at the international fair.

One of the first things you notice is that some categories have a lot more projects than others, especially in the younger grades. I think that the most important reason for this relates to the science classes the students are currently taking. Another reason is that in certain categories, more possibilities exist for projects that directly relate to the students' lives.

Applying the Acid Test: Can I Really Do This Project?

Some students have huge, noble objectives for their science projects, but to be successful, you must have a project that you can research, plan, and execute, according to your abilities and the time available. This section helps you take a critical look at your project idea to see whether it fits.

Is it within my abilities?

One mistake that students make is biting off more than they can chew. Remember that as much as you want to, you're not going to find a cure for the common cold doing a four-month eighth grade science project. You may, however, find one small aspect of that topic that you can do — for example, testing which over-the-counter cold medicine kills the most bacteria.

Ask whether you have enough knowledge and experience to do this particular project. Have you covered any of this material in your science, math, or computer classes? Have you done anything like this before? A "no" answer to these questions doesn't automatically rule out your idea, but if you find butterflies fluttering in your stomach at the very thought of doing the project, it probably isn't going to work. Keep in mind that you must prepare a presentation for your project, so you want to make sure that you know your topic well enough to be able to explain it in detail and answer questions about it.

One more thing: Even if you have an expert in your family, remember that this is your project, not your Mom's or Dad's.

Do I have enough time?

Usually, you have three to four months to do your entire science project, and that includes doing your research and preparing the final lab report and project display. Figure out if the project has anything you need that either takes more time than you have or relies on factors outside your control.

For example, a project that requires a certain amount of rainfall may be unsuccessful in a drought year, and a botany experiment that uses a slow-growing shrub may require more time than you have.

Can I afford it?

A successful project doesn't require a lot of money. In fact, some very complex and successful projects have been done for very little money.

But I do recommend figuring out how much money you need to do your work. Before you make a final decision on a project idea, figure out how much everything's going to cost, from the materials for your experiment to the art supplies for your display. To help you estimate your project costs, make a budget that includes as many things as possible.

Estimating your budget may get you thinking about how to get what you need without buying it (and without stealing it!). You don't need access to a million-dollar computer or a state-of-the art laboratory to do a thorough and creative experiment. For example, you may use the equipment on high school and university campuses when classes aren't in session. Using such equipment lets you do a project that you otherwise may not be able to afford.

Also, teachers may be able to furnish you with some items or to order certain supplies. (Some items can be shipped only to schools.)

Do my parents and teacher approve?

Because you may be doing a great deal of your project at home, make sure that your idea is okay with your parents by asking them some questions. For example:

- ✔ Do you mind that the guest room is full of my stuff for the next three months?
- ✔ Can I keep the E. coli bacteria in the food containers?
- ✔ Is it okay to let lettuce rot in the refrigerator?
- ✔ Do you mind having jars of worms, ants, and fruit flies in the bathroom (or, worse yet, the kitchen)?

If you're required to do a science project, your teacher likely wants to see a statement of your project idea by a certain date. He or she looks at whether your idea is feasible and whether it can be considered a true science project. For example, if your project idea is simply to build a model without doing any testing or analysis, your teacher may veto it and send you back to the drawing board.

 If you look at science fair programs from previous years, you notice that certain project categories are very popular, and certain projects are often repeated. You don't necessarily need to find something completely original, but if you want to use an idea that's been done often, try to find a new twist — remember that teachers and judges get bored, too!

Is it safe?

If you're planning to use chemicals or electricity in your project, you need to make sure that you can use them safely. Ask your parents and teacher to help you figure out what issues you need to consider so you aren't tackling anything that may burn your house down or put anyone in danger.

Does my project follow the rules?

The Society for Science & the Public (SSP), which runs the Intel International Science and Engineering Fair (ISEF), has developed a list of rules that help to ensure the safety of students and the well-being of any specimens used in experiments. There are rules governing the use of live vertebrates, for example, as well as controlled substances, chemicals, bacteria, and tissue samples. To get a sense of the rules, go to www.societyforscience.org/isef and click on Get Started. Look for a link to Rules & Guidelines.

Chapter 4

Starting Out with an Easy Project

In This Chapter
- Analyzing human behavior
- Planting seeds: Botany projects
- Studying what can't be seen: Microbiology projects

*I*n this chapter, I take you on a tour of some biology science fair projects that demonstrate how you can take ordinary things that you use every day and create a project that's out of the ordinary!

Each project described here was actually displayed at the Greater San Diego Science and Engineering Fair, which means that these projects were selected as the best projects from their school science fairs.

Behavioral and Social Sciences

Most projects in this category utilize qualitative analysis. To do them, you collect data from human subjects and analyze it to prove the hypothesis. Following is an example.

Does age affect the ability to remember?

Bonnie Carr believed that older people had worse memories than younger people. She decided to test how age affects your ability to remember.

Hypothesis
Age negatively impacts the ability to remember.

Experimental variables
Age groups whose memory will be tested

Measured variables
The number of correct answers that each subject gives

Controls
The memory test given to each subject

Experimental groups
- 10–19 year olds
- 20–39 year olds
- 40–59 year olds
- 60–79 year olds

Materials
- Four game cards from Stare game
- Timer
- Four groups of five participants

Procedures
1. **Show the group a Stare card for 30 seconds.**
2. **Hide the card.**

3. **Have each participant answer five questions about the card shown.**
4. **Record the number of correct answers.**
5. **Repeat the procedure for the other groups.**

Results

- 10–19 year olds: 56 percent correct
- 20–39 year olds: 47 percent correct
- 40–59 year olds: 55 percent correct
- 60–79 year olds: 48 percent correct

Conclusions

My hypothesis that age would negatively affect memory was incorrect. I found only very slight differences in the number of correct answers by age group. One possible reason can be that adults were more familiar with the objects that I used. Altering the age ranges may improve my project, because the first range was 10 years, and all other ranges were 20 years.

Botany

Many students choose botany projects because they're fairly easy to set up. Following is an example.

Deterring whitefly on home gardens

Some of the things that people use to kill pests can be toxic to the environment. Mitzi Larson thought that a common household substance could kill garden pests without bad effects on the environment.

Hypothesis
Palmolive dish soap will do the best job of deterring whiteflies.

Experimental variables
Substances used to control whiteflies

Measured variables
Number of whiteflies after treatment

Controls
- ✔ Soil
- ✔ Types of plants
- ✔ Plant containers

Experimental groups
- ✔ Plants sprayed with acetone
- ✔ Plants sprayed with corn oil
- ✔ Plants sprayed with Palmolive soap
- ✔ Plants sprayed with WD-40
- ✔ Plants sprayed with coffee

Control groups
Plants sprayed with water

Materials
- ✔ Water
- ✔ Acetone
- ✔ Corn oil
- ✔ Palmolive soap

- WD-40
- Coffee
- 24 casaba melon seeds
- 24 disposable foam cups
- Dirt
- Potting soil
- Six spray bottles

Procedures

1. **Set up experimental and control groups:**
   a. Punch eight holes into each disposable foam cup.
   b. Mix equal amounts dirt and potting soil.
   c. Put 3.5 cups dirt/potting soil mixture into each cup.
   d. Put a melon seed in each cup.
   e. Cover seed with .5-cup dirt/soil mix.
2. **Separate plants into six groups of four plants each, labeled with the name of the deterrent to be used.**
3. **After plants grow, introduce whiteflies.**
4. **When 10 to 15 whiteflies appear on each leaf, spray each group with equal amounts of the deterrent.**
5. **In three days, count and record the number of whiteflies.**
6. **Repeat Steps 4 and 5 three more times.**

Results
Table 4-1 shows the whiteflies counted every third day.

Table 4-1	Number of Whiteflies			
	Check 1	*Check 2*	*Check 3*	*Check 4*
Water	10	27	23	18
Acetone	7	12	12	15
Corn oil	9	18	20	14
Palmolive	3	4	4	5
WD-40	6	16	15	17
Coffee	8	18	19	15

Conclusions
The hypothesis that Palmolive dish soap would do the best job of deterring whiteflies was correct because the group treated with Palmolive had the fewest whiteflies.

Microbiology

Microbiology projects look at the world of things that you can't see — *microorganisms,* which are organisms of microscopic or submicroscopic size, such as bacteria. Following is an easy example of this type of project.

What's the weakest solution of bleach that will kill bacteria?

Washing dishes in diluted bleach will kill germs. But because bleach can also have some negative effects,

Jaleesa Chavez wanted to find out the least amount of bleach that was needed to kill bacteria. She wondered what the smallest amount of bleach is that's needed to kill E. coli bacteria.

Hypothesis
The smallest strength of bleach needed to kill E. coli bacteria will be a 3 percent solution.

Experimental variables
Strength of bleach solution applied to E. coli bacteria

Measured variables
Percent of E. coli killed by bleach solution

Controls
Amount of solution applied to E. coli bacteria

Experimental groups
- 3 percent bleach solution
- 5 percent bleach solution
- 8 percent bleach solution
- 10 percent bleach solution
- 15 percent bleach solution

Control groups
Water

Materials
- Bleach
- Water
- E. coli bacteria

Procedures

1. **Create solutions of 3 percent, 5 percent, 8 percent, 10 percent, and 15 percent bleach in water.**
2. **Spray each solution on E. coli bacteria.**
3. **Measure amount and percentage of E. coli bacteria killed by the bleach solution.**

Results

- The 3 percent bleach solution killed 1.75 percent of the bacteria.
- The 5 percent bleach solution killed 2.4 percent.
- The 8 percent bleach solution killed 2.6 percent.
- The 10 percent bleach solution killed 2.7 percent.
- The 15 percent bleach solution killed 4.1 percent.

Conclusions

My hypothesis stated the smallest concentration of bleach in water that would kill E. coli bacteria. Of the solutions tested, the 3 percent solution was the smallest concentration that will kill bacteria, although the amount of germs killed went up in proportion to the concentration of bleach.

Chapter 5

Kicking It Up a Notch: Medium Difficulty Projects

In This Chapter
- Studying right- and left-handed behavior
- Testing antibacterial effectiveness

This chapter features two ideas for slightly more advanced biology projects. These projects were displayed at the Greater San Diego Science and Engineering Fair, which means that they were selected as the best from their school science fairs.

The Even-Handed Teacher

Are teachers fair? Who knows? Are they even-handed? That can be tested, and Lea Cohen did just that.

Hypothesis

A right-handed teacher pays more attention to the left side of the room, so that he or she can pay attention to the class while writing on the board.

Independent variables

Left- and right-handed teachers

Dependent variables
Classroom location of students called on

Experimental groups
Fourteen classrooms, between fourth and eighth grades

Procedures
Visit 14 classrooms four times, for ten minutes each, and observe the following:

- Whether the teacher is left- or right-handed
- The arrangement of the classroom (in rows, in a semi-circle, and so on)
- The location and frequency of the students called on

Results
On average, the right-handed teachers called on students:

- On left side of room: 3.17 times
- In center of room: 1.69 times
- On right side of room: 2.88 times

Note: Only one left-handed teacher was in the sample.

Conclusions
My hypothesis, that a right-handed teacher would pay attention to the left side of the room so that he or she can pay attention to the class while writing on the board, was correct.

A possible improvement in my project would be to include more left-handed teachers in the test.

Hand Sanitizers: Hype or Help?

Many soaps and cleansers claim to be antibacterial. Is it true? Steven Paletz wondered whether ethanol is the only effective waterless antibacterial agent.

Hypothesis
Waterless hand sanitizers and generic ethanol are equally effective in killing bacteria. AloeGuard, which contains chloroxylenol, will be just as effective as ethanol products, because it is used in hospitals.

Independent variables
Antibacterial cleaners

Dependent variables
Number of bacteria colonies

Controls
- Bacteria tested
- Incubation type and period
- Measurement methods

Experimental groups
- Purell
- AloeGuard
- Alcare Plus
- Bath & Body Works antibacterial hand sanitizer
- 62 percent solution of ethyl alcohol
- 6.2 percent solution of ethyl alcohol

Control groups
No cleaner used

Materials
- One 8 oz. bottle of Purell waterless hand sanitizer with aloe
- One 8 oz. bottle of Bath & Body Works waterless hand sanitizer
- One 9 oz. can of Alcare Plus waterless hand wash foam
- One 4 oz. bottle of AloeGuard waterless hand cleansing lotion
- 40 petri dishes with nutrient *agar* (a seaweed product used as a medium for growing cultures)
- Ethyl alcohol
- Rotating plate
- Bunsen burner
- Glass rod
- XL1 blue strain of E. coli bacteria
- 40 test tubes
- Eyedropper
- Vortex
- Incubator
- Gloves
- Distilled water
- Tape
- Pipette

Procedures
1. **Put on gloves and other protection.**
2. **Dilute the different hand sanitizers in the bacteria (using the testing tube) as follows:**

a. For the 1:1 dilution, 100 ml of bacteria to 100 ml of the cleanser

b. For the 2:1 dilution, 50 ml of bacteria to 100 ml of the cleanser

3. **Shake the tubes that contain bacteria and hand sanitizers and let them sit for 15 minutes.**
4. **Place 50 ml of each dilution onto a petri dish.**
5. **Place the bacteria in the incubator for 24 hours at 37° Celsius.**
6. **After incubation period, count the bacteria colonies and record the results.**
7. **Seal the petri dishes with tape and dispose of them properly.**
8. **Repeat for trials 2 and 3.**

Results

Table 5-1 shows the number of colonies remaining after 24 hours, for both the 1:1 and the 2:1 concentration of each antibacterial.

Table 5-1			Bacteria Remaining				
Trial	Purell	Aloe-Guard	Alcare Plus	Bath & Body Works	6.2% Ethyl Alcohol	Control	62% Ethyl Alcohol
Trial 1 1:1	>1,000	0	>1,000	>1,000	>1,000	>1,000	0
Trial 1 2:1	250	0	250	340	355	360	0
Trial 2 1:1	>1,000	0	>1,000	>1,000	>1,000	>1,000	1

(continued)

Table 5-1 *(continued)*

Trial	Purell	Aloe-Guard	Alcare Plus	Bath & Body Works	6.2% Ethyl Alcohol	Control	62% Ethyl Alcohol
Trial 2 2:1	460	0	240	300	250	380	0
Trial 3 1:1	>1,000	0	>1,000	>1,000	>1,000	>1,000	0
Trial 3 2:1	400	0	255	370	490	240	0

Table 5-2 shows the average number of colonies, for both the 1:1 and the 2:1 concentrations of each antibacterial.

Table 5-2   Average Amount of Bacteria Remaining for Each Concentration

	Purell	Aloe-Guard	Alcare Plus	Bath & Body Works	6.2% Ethyl Alcohol	Control	62% Ethyl Alcohol
1:1	>1000	0	>1,000	>1,000	>1,000	>1,000	0.33
2:1	370	0	248	337	365	327	0.00

Conclusions

My hypothesis that waterless hand sanitizers, generic ethanol, and AloeGuard are equally effective at killing bacteria wasn't correct. Testing showed that the antibacterial hand sanitizers didn't live up to their claims. However, the chloroxylenol found in AloeGuard was very effective, killing 100 percent of the bacteria colonies tested after 24 hours.

Chapter 6

Taking the Challenge: Harder Projects

In This Chapter

▶ Testing bacteria levels in prepackaged lettuce
▶ Walking through a more complex project procedure

If you've had some previous experience with science or science fair projects, you may want to try something that's a little more challenging. This chapter shows you a sample microbiology project and, I hope, offers some inspiration. Germs are everywhere, and most people are concerned about keeping them from contaminating the world. Therefore, microbiology projects are common sights at science fairs.

This project was actually displayed at the Greater San Diego Science and Engineering Fair, which means that it was selected as the best project from a school science fair.

To Clean or Not to Clean

If you eat ready-to-eat packaged lettuce, does it contain bacteria? And if so, how can you wash it to most effectively get rid of the bacteria? Jane Alejandro really wanted to know.

Hypothesis
Washing pre-washed lettuce with a vinegar and water solution will remove more bacteria than washing with water only.

Independent variables
Type of wash used for lettuce

Dependent variables
Amount of bacteria on samples after three-day test

Controls
- Lettuce samples
- Sterilization techniques

Experimental groups
- Lettuce washed with plain tap water
- Lettuce washed with mixture of vinegar and water

Control groups
Lettuce that wasn't washed

Materials

- 37 nutrient agar plates (which contain a culture to grow bacteria)
- 40 sterile plates
- Six packages pre-washed lettuce
- 70 percent ethanol
- De-ionized water
- Two sets of tweezers
- Two 100-ml beakers
- One 200-ml beaker
- One Bunsen burner
- 45 sterile cotton balls
- One liter of tap water
- One liter of vinegar (bottled)
- Three stirring rods

Procedures

To perform this experiment, take a piece of lettuce from each package, swipe it with a cotton swab, and streak the swab on the nutrient agar plate. Then place the agar plates in an incubator set at 37° C.

When bacterial growth is evident, take the same pieces of lettuce and put them through either a tap water or vinegar/water wash. Note that on each day, the workstation, tweezers, and stirring rods are cleaned and sterilized with 70 percent ethanol.

The following is a closer look at the procedures.

✔ Day 1: Incubate bacterial growth:

1. Mark each lettuce sample as XI, XII, YI, YII, ZI and ZII.
2. Label one nutrient agar plate the control plate and streak with de-ionized water.
3. Divide 18 nutrient agar plates into six groups of three. Label the first group XI, and assign a number to each plate (for example, XI 1, XI 2, and XI 3). Repeat with remaining groups (XII, YI, and so on).
4. Dampen the cotton ball with the de-ionized water and streak it twice across the agar on the control plate.
5. Place a piece of lettuce on each labeled sterile nutrient agar plate.
6. Incubate nutrient agar plates overnight at 37° C.
7. Refrigerate sterile plate with lettuce piece at 4° C.

✔ Day 2: Wash lettuce with tap water or vinegar and water:

1. Observe, note, illustrate, and photograph the agar plates.
2. Divide the remaining 18 nutrient agar plates into six groups of three. In each group, label one as the control plate, the second PW (plain water), and the third group VW (vinegar/water mix).

3. Group each sterile plate with its corresponding nutrient agar plate containing a lettuce sample.
4. Bathe PW samples in tap water using stirring rod.
5. Bathe VW samples in vinegar-water mix using stirring rod.
6. Incubate the nutrient agar plates overnight at 37° C.

Day 3: Observe and record bacterial growth:

1. Remove agar plates from incubator.
2. Observe, note, illustrate, and photograph the agar plates.

Results

All samples washed with tap water grew bacteria, while only one sample washed with vinegar mix grew bacteria.

On day 2, after incubating the bacteria overnight, each plate had bacterial growth except for the control plate. Many distinct colonies formed, with two massive colonies in one of every group. All brands had relatively equal bacterial growth.

On day 3, all the control plates had bacterial growth. Only one of the VW plates had bacterial colonies formed, but all the PW plates had evident bacterial growth.

None of the plates were clear of bacterial growth.

Conclusions

My hypothesis was that re-washing pre-washed lettuce with a vinegar and water solution removes more bacteria than re-washing with water. The results show that this hypothesis is correct.

The significance of my findings is that bacteria does exist in packaged, ready-to-eat lettuce despite its label, which says "no washing needed." The best way to get rid of the bacteria is to wash the lettuce with a mixture of vinegar and water.

More Biology Challenges

If sanitizing lettuce isn't your thing, you have a world full of biology challenges that could make for great science fair projects. Here are just a couple more ideas to get you thinking:

- Does garlic keep vampires away? You can't prove it scientifically, but many people believe that garlic is good for what ails you. Among other things, it helps to ward off infection. If you've ever noticed that people who eat garlic tend to have fewer colds than others, you may develop a hypothesis and testing procedure to see if there's any truth in that idea.

- With so many acne cleansers and medications available, how do you know which ones are actually worth the money? Do some work better than others? And if so, does the price tag usually correspond with product performance?

Of course, the world of biology projects can involve growing plants, testing animal behavior, and much more. Just take a look around your environment, and chances are you can generate a challenging project idea without too much stress.

Chapter 7
Ten Survival Tips for Parents

In This Chapter
- Having fun
- Remembering whose project it really is

Very often, parents are just as stressed as their sons and daughters at science project time, if not more so. This chapter takes a look at ten ways to make the science project journey pleasant and fun for everyone.

Supporting Versus Nagging

If you've ever had a nagging parent, spouse, or boss, you know the difference between supporting and nagging. Instead of getting both you and your child worked up, try the following:

- Give your child a convenient, visible place to post a calendar to keep track of his or her own assignments and deadlines.
- Allow your child to make a mistake or two.
- No matter how close the deadline or how much still needs to be done, allow your child some down time to rest and recharge.

Discovering New Things

If you haven't studied much science over the years, you may be intimidated by the thought of helping your kids with their science projects. However, this can be a great opportunity to find out about some things you were always curious about or didn't even know existed.

Making Friends with Your Child

One of the best things about doing a science project is the opportunity to spend some quality time together and develop common interests. While preparing for the presentation, act as an audience for your child and give suggestions on how to make communication better.

Living Your Second Childhood

Your child's science project may be a great opportunity to shoot off a bottle rocket, play with a chemistry set, or watch insects reproduce. How many other times do you get to do that just for fun?

And if you enjoyed arts and crafts as a kid, you can really have fun when it's time to build the project display.

Knowing When to Say No

How much help you can and will give is entirely up to you. For example, helping your child order supplies online may be perfectly reasonable. However, running out to the store at 9 p.m. when your child knew what she needed three days ago may be over the line.

My best advice is to feel free to set reasonable limits. You have the right to refuse to respond to any request at any time, set a spending limit, and even object to a colony of earthworms in your bathroom sink.

The best way to deal with conflict is to prevent it. Make your expectations clear at the beginning of the project. Then, stick to your guns and just say no!

Taking Time for Yourself

No matter what your child needs, remember that his or her science project is no reason to put your life on hold.

You don't need to call off your golf game, stop going to yoga class, or cancel an evening out with friends. You do a lot for your family, and probably more while your child is working on a science project.

Staying Centered

Take a minute to get calm and quiet, and the confusion that a project can often bring won't seem so overwhelming. Remember that your child knows what to do and has been instructed. Help ensure that your child is staying calm by remaining calm yourself.

Getting a Self-Checkup

Ask yourself how your life is going (and not just as a parent, either).

On a scale of 1 to 10, rate yourself in relationships, friends and family, home environment, career, and personal growth and fulfillment. Compare your results each week. If things are going downhill during science project time, it may be time to adjust your priorities.

Letting It Go

Whose science project is it anyway? Don't get so preoccupied that the project becomes more important to you than to your child. Keep reminding yourself that her grades or awards belong to her and not to you.

You can help, but don't do the project yourself — you're not doing your child any favors. Besides, teachers and judges usually know when a child didn't do the work.

Asking for Help

If you need assistance with the project itself, look for some expert advice online, reach out to friends who can help, contact the folks who run the science fair, or ask your child's teacher for advice.

If you're really lucky, you may have an older son or daughter who's already done a science project.

Apple & Macs

iPad For Dummies,
2nd Edition
978-1-118-02444-7

iPhone For Dummies,
5th Edition
978-1-118-03671-6

iPod touch For Dummies,
3rd Edition
978-1-118-12960-9

Mac OS X Lion
For Dummies
978-1-118-02205-4

Blogging & Social Media

CityVille For Dummies
978-1-118-08337-6

Facebook For Dummies,
4th Edition
978-1-118-09562-1

Mom Blogging
For Dummies
978-1-118-03843-7

Twitter For Dummies,
2nd Edition
978-0-470-76879-2

WordPress
For Dummies,
4th Edition
978-1-118-07342-1

Business

Cash Flow For Dummies
978-1-118-01850-7

Investing For Dummies,
6th Edition
978-0-470-90545-6

Job Searching
with Social Media
For Dummies
978-0-470-93072-4

QuickBooks 2011
For Dummies
978-0-470-64649-6

Resumes For Dummies,
6th Edition
978-0-470-87361-8

Starting an Etsy Business
For Dummies
978-0-470-93067-0

Cooking & Entertaining

Cooking Basics
For Dummies, 4th Edition
978-0-470-91388-8

Wine For Dummies,
4th Edition
978-0-470-04579-4

Diet & Nutrition

Kettlebells For Dummies
978-0-470-59929-7

Nutrition For Dummies,
5th Edition
978-0-470-93231-5

Restaurant Calorie
Counter For Dummies,
2nd Edition
978-0-470-64405-8

Digital Photography

Digital SLR Cameras &
Photography
For Dummies, 4th Edition
978-1-118-14489-3

Digital SLR Settings
& Shortcuts
For Dummies
978-0-470-91763-3

Photoshop Elements 9
For Dummies
978-0-470-87872-9

Gardening

Gardening Basics
For Dummies
978-0-470-03749-2

Vegetable Gardening
For Dummies,
2nd Edition
978-0-470-49870-5

Green/Sustainable

Raising Chickens
For Dummies
978-0-470-46544-8

Green Cleaning
For Dummies
978-0-470-39106-8

Health

Diabetes
For Dummies,
3rd Edition
978-0-470-27086-8

Food Allergies
For Dummies
978-0-470-09584-3

Living Gluten-Free
For Dummies,
2nd Edition
978-0-470-58589-4

Hobbies

Beekeeping
For Dummies,
2nd Edition
978-0-470-43065-1

Chess For Dummies,
3rd Edition
978-1-118-01695-4

Drawing For Dummies,
2nd Edition
978-0-470-61842-4

eBay For Dummies,
7th Edition
978-1-118-09806-6

Knitting
For Dummies,
2nd Edition
978-0-470-28747-7

Language &
Foreign Language

English Grammar
For Dummies,
2nd Edition
978-0-470-54664-2

French For Dummies,
2nd Edition
978-1-118-00464-7

German For Dummies,
2nd Edition
978-0-470-90101-4

Spanish Essentials
For Dummies
978-0-470-63751-7

Spanish For Dummies,
2nd Edition
978-0-470-87855-2

Available wherever books are sold. For more information or to order direct: U.S. customers visit
www.dummies.com or call 1-877-762-2974. U.K. customers visit www.wileyeurope.com or
call (0) 1243 843291. Canadian customers visit www.wiley.ca or call 1-800-567-4797.
Connect with us online at www.facebook.com/fordummies or @fordummies

Math & Science

Algebra I For Dummies,
2nd Edition
978-0-470-55964-2

Biology For Dummies,
2nd Edition
978-0-470-59875-7

Chemistry For Dummies,
2nd Edition
978-1-1180-0730-3

Geometry For Dummies,
2nd Edition
978-0-470-08946-0

Pre-Algebra Essentials
For Dummies
978-0-470-61838-7

Microsoft Office

Excel 2010 For Dummies
978-0-470-48953-6

Office 2010 All-in-One
For Dummies
978-0-470-49748-7

Office 2011 for Mac
For Dummies
978-0-470-87869-9

Word 2010
For Dummies
978-0-470-48772-3

Music

Guitar For Dummies,
2nd Edition
978-0-7645-9904-0

Clarinet For Dummies
978-0-470-58477-4

iPod & iTunes
For Dummies, 9th Edition
978-1-118-13060-5

Pets

Cats For Dummies,
2nd Edition
978-0-7645-5275-5

Dogs All-in-One
For Dummies
978-0470-52978-2

Saltwater Aquariums
For Dummies,
2nd Edition
978-0-470-06805-2

Religion & Inspiration

The Bible For Dummies
978-0-7645-5296-0

Catholicism
For Dummies,
2nd Edition
978-1-118-07778-8

Spirituality For Dummies,
2nd Edition
978-0-470-19142-2

Self-Help & Relationships

Happiness For Dummies
978-0-470-28171-0

Overcoming Anxiety
For Dummies,
2nd Edition
978-0-470-57441-6

Seniors

Crosswords For Seniors
For Dummies
978-0-470-49157-7

iPad For Seniors
For Dummies, 2nd Edition
978-1-118-03827-7

Laptops & Tablets
For Seniors
For Dummies,
2nd Edition
978-1-118-09596-6

Smartphones & Tablets

BlackBerry
For Dummies, 5th Edition
978-1-118-10035-6

Droid X2 For Dummies
978-1-118-14864-8

HTC ThunderBolt
For Dummies
978-1-118-07601-9

MOTOROLA XOOM
For Dummies
978-1-118-08835-7

Sports

Basketball For Dummies,
3rd Edition
978-1-118-07374-2

Football For Dummies,
4th Edition
978-1-118-01261-1

Golf For Dummies,
4th Edition
978-0-470-88279-5

Test Prep

ACT For Dummies,
5th Edition
978-1-118-01259-8

ASVAB For Dummies,
3rd Edition
978-0-470-63760-9

The GRE Test
For Dummies, 7th Edition
978-0-470-00919-2

Police Officer Exam
For Dummies
978-0-470-88724-0

Series 7 Exam
For Dummies
978-0-470-09932-2

Web Development

HTML, CSS, & XHTML
For Dummies, 7th Edition
978-0-470-91659-9

Drupal For Dummies,
2nd Edition
978-1-118-08348-2

Windows 7

Windows 7
For Dummies
978-0-470-49743-2

Windows 7
For Dummies,
Book + DVD Bundle
978-0-470-52398-8

Windows 7 All-in-One
For Dummies
978-0-470-48763-1

Available wherever books are sold. For more information or to order direct: U.S. customers visit www.dummies.com or call 1-877-762-2974. U.K. customers visit www.wileyeurope.com or call (0) 1243 843291. Canadian customers visit www.wiley.ca or call 1-800-567-4797.
Connect with us online at www.facebook.com/fordummies or @fordummies

| In store only | Expires June 29, 2013 |

$1 OFF

any ArtSkills poster making product

Valid through 6/29/13 in Staples® U.S. stores only. Consumer: Limit one offer per coupon. Only valid on ArtSkills items. Cannot be combined with the same or any other offer. Only valid at Staples stores. Void if copied, scanned, transferred, sold or prohibited by law. Duplicated or altered coupons will not be accepted. May not be valid at all locations. One time use.

71987

STAPLES